AUSTRALIA'S DANGEROUS WILDLIFE

Sydney funnelweb spider (*Atrax robustus*), see page 35.

(*Left*) Estuarine stonefish (*Synanceia trachynis*), see page 19.

Text by Dalys Newman

AUSTRALIAN PICTURE LIBRARY

CHILD & ASSOCIATES
AN ALL-AUSTRALIAN PUBLISHER

Acknowledgements

The publishers wish to acknowledge the valuable assistance of the Australian Picture Library staff and the people whose material is held by the Australian Picture Library.

Robert Berthold pages 4, 18, 35 (below)
Ray Bisson page 31 (below)
John Carnemolla pages 14, 17, 20 (below), 21, 23 (above), 30 (below), 34 (above), 39 (above), 43, 48 (above and centre)
Rob Cooper 6, 62–3
Dive 2000 32, 33
Ron Dorman 23 (below)
Colin Driscoll 20 (above), 38, 41 (left)
Roger Garwood 12
Ken Griffiths back cover, pages 2, 3, 5, 11, 35 (above), 36 (above), 42, 44 (above), 46, (above), 47 (above), 48 (below), 49, 51, 52 (below), 53, 54, 55, 56, 57, 58 (below), 59, 60 (above)
Leigh Hemmings page 41 (right)
Raymond Hoser pages 31 (above), 45, 50 (below), 52 (above), 58 (above)
Tony Joyce 44 (below), 60 (below)
Noeline Kelly page 10
Gary Lewis page 27
Jim McKay page 24, 25
Outback Photography pages 13, 37 (below)
Fritz Prenzel page 46 (below)
Norman Quinn page 16 (below)
Derek Roff pages 15, 47 (below), 50 (above)
Lindon Schick pages 34 (below), 36 (above)
A. Tolhurst pages 37 (above), 39 (below)
Volvox pages 8, 9, 26
Peter Wilson page 40
Bill Wood front cover, pages 10, 19, 22, 28, 29, 30 (above)

The author gratefully acknowledges the following people for their assistance in identification of species:
Tim Hawkes, Reptile Keeper, Taronga Park Zoo.
Stephen Menzies, B.Sc., Australia's Underwater World.
John Riley, B.Sc., entomologist.

Above: Saltwater or estuarine crocodile *(Crocodylus porosus)*, see page 60.
Front cover: Leopard moray eel *(Gymnothorax flavimarginatus)*, see page 22.
Back cover: Eastern brown snake *(Pseudonaja textilis)*, see page 48.
Endpaper: Collett's snake (*Pseudechis collettii*), see page 55.

Published by
Child & Associates Publishing Pty Ltd
5 Skyline Place, Frenchs Forest, NSW, Australia, 2086
A wholly owned Australian publishing company
This book has been edited, designed and typeset in Australia by the Publisher

First edition 1989
Text by Dalys Newman
Photographs from the Australian Picture Library

Printed in Singapore by Kyodo-Shing Loong Printing Industries Pte Ltd
Typesetting processed by Deblaere Typesetting Pty Ltd

National Library of Australia Cataloguing-in-Publication data

Newman, Dalys.
Australia's dangerous wildlife.

Includes index.
ISBN 0 86777 134 8.

1. Dangerous animals—Australia. 2. Dangerous animals—Australia—Pictorial works. I. Australian Picture Library. II. Title.

591.6'5'0994

INTRODUCTION

Australia and its coastal waters abound with creatures that are capable of harming or even killing man. From the time man first inhabited this continent he has learned to live alongside these creatures; folklore, legend and superstitions have grown up around them, but it is only recently that man has fully understood the complexities of these fascinating animals. With this understanding has grown a respect for the biological richness that they add to our unique wildlife.

Dangerous creatures are widespread over the entire continent and their distribution patterns often take them into inhabited areas: the popular tropical waters of the northern coral coast contain the lethal sea wasp; harbour beaches and reefs around Sydney and Melbourne are home to the blue-ringed octopus; tiger snakes patrol the outskirts of our southern cities; taipans are found in inhabited northern areas and funnelweb spiders are residents of Sydney's most exclusive suburbs.

Despite the wide variety and distribution of these venomous and dangerous animals, nowadays there are relatively few fatalities. This is due largely to the development of antivenoms, along with the fact that most of these animals are shy; 'attacks' that do take place being mainly accidental. None of the infinite killing mechanisms employed by animals is specifically meant for use against humans. Most of the venomous animals use their poison as a hunting weapon to immobilise potential food—their contact with man and the death that may result is by sheer chance. Even encounters with the infamous man-eaters—crocodiles and sharks—are usually the result of inviting trouble. However, in order to minimise a chance encounter with one of the deadly Australians, it is wise to know and understand the distribution and behavioural patterns of these fearsome and fascinating creatures. Most fatalities occur through carelessness and sheer ignorance and could be avoided if we learn to co-exist peacefully with potentially harmful wildlife.

Marine Animals

Australian waters contain an incredible diversity of marine fauna, with many different fishes capable of harming man in some way. These creatures, over a long evolutionary period, have developed a considerable variety of defensive and offensive structures.

Marine animals that are potentially dangerous to man can be divided into three groups: those that bite; those that are poisonous to eat, and those that sting.

Marine creatures that bite

Among the biting animals, sharks have no rivals when it comes to instilling terror. One of the most primitive fears is that of being eaten alive by one of these hungry and undiscriminating carnivores. According to fossil evidence, sharks have been in the seas for more than three million years and about three hundred species have been identified to date. Most are found in tropical waters, but many, such as the white pointer, prefer cooler seas. Species found in Australian waters include the black tip shark, graceful or school sharks, grey nurse, whalers, tiger shark, hammerhead and white pointer. The mako, or blue pointer, a very aggressive and dangerous shark, is found in tropical and warm-temperate Australian seas, but rarely enters nearshore water where most people swim. Bluewater boat anglers often encounter the mako, which, when hooked, is notorious for its attacks on small crafts.

Sharks differ from bony fish in several ways. Their skeletons are made up of cartilage rather than bones, and instead of scales the skin has thousands of teeth—tiny sharp points covered with hard enamel. And, unlike most fish, they must be perpetually on the move in order to breathe and stay afloat. Most fish have air bladders inside their bodies that make them lighter than water. Sharks do not and therefore must swim continually otherwise they will sink. In some species, however, the liver is large and full of oil (a substance lighter than water) and this aids their flotation. Bony fish use their mouths as pumps, forcing water back to their gills, enabling them to breathe. Most sharks can't pump water in this manner, instead they keep swimming,with their mouths open, to force water into their gills.

Many attempts have been made to define the pattern of shark attack and several broad scientific conclusions have been reached. One of the most clearly established factors is water temperature; attacks occur almost exclusively when the water temperature is 21°C or more. (An exception is the white pointer which has attacked in waters of 13°C or less.) In Australia, the tropical zone, where attacks occur all year round, covers the northern coastline from Port Hedland, Western Australia, to Mackay, Queensland. With the onset of summer, warm currents move down the western and eastern coasts, and by January, the boundary of the attack zone reaches Bass Strait—so encompassing the entire continental coastline.

Another known factor in shark attack is the animal's keen acoustic senses. They get a very rapid and accurate fix on the movements of likely prey. Thus the movement of a splashing swimmer (similar to that of a struggling marine animal) will send out pressure waves which the shark detects by a line of highly sensitive cells along both sides of its body.

All sharks are carnivorous, their preference being live fresh prey, although some will scavenge decaying flesh. The diets of different species range from microscopic plankton to large seals and whales.

Another notorious underwater biter is the marine eel. Most have sharp, fanglike teeth and are aggressively defensive if attacked or disturbed in the holes in the reef where they live.

A mako shark (*Isurus glaucus*), one of the most fearsome of oceanic creatures. This species appears in abundance off the New South Wales coast even in winter, but rarely enters near shore waters.

Moray eels can inflict serious wounds, but their bite is not venomous, the danger from the wound lies in infection.

Marine creatures that are poisonous to eat

The risk of poisoning from eating fish is another danger concerning marine animals. Some species, such as puffers, always contain certain toxic substances in their organs. There are about thirty species of pufferfish (including porcupinefish and toadfish) in Australian coastal waters, estuaries and even tidal creeks. A number of fatalities have occurred from eating these poisonous fish. The toxin responsible for death is called tetradoxin and is a nerve poison. In serious cases, major symptoms develop within ten to forty-five minutes after eating the fish. There is numbness, and muscle activity and heartrate are reduced causing lowered blood pressure. Victims may remain conscious but cannot swallow, see, speak or move. Breathing failure is the major life-threatening effect. In Japan, where specially prepared pufferfish are a delicacy, the death rate for untreated cases is 60 per cent. To date no antidote has been found.

Fish containing the toxin vary markedly in body form and colour; what they all have in common is a lack of true scales, instead they are covered by a defensive bumpy plating. In porcupinefish the body scales are modified into spines, usually lying flat, but when the fish is inflated they stick out at right angles.

Marine creatures that sting

A much more serious hazard than inedible fish is encountered in several groups of venomous fish which possess poison glands and devices for injecting this poison into their victims. One such family is the Scorpaenidae, or scorpionfish. An attractive group of this family, known variously as butterfly cod, firefish, zebrafish, turkeyfish, lionfish and featherfins, have well-camouflaged spines in their beautiful lacy fins. Each spine has two narrow grooves, extending partly along the shaft. A venom gland is located in each groove. If these fish are touched the spines may penetrate the skin, injecting venom. Immediate and intense shooting pain results from a sting and collapse from shock can sometimes follow. Another group of scorpionfish, including the somewhat less flamboyant cobbler, fortescue and bullrout, have shorter, stouter spines with a slightly less effective venom apparatus. Their sting is painful but is rarely serious.

Closely related to the scorpionfish, the stonefish is probably the best known of all Australia's poisonous fishes. They are so well camouflaged that they are almost always mistaken for a piece of algae-covered rock. If inadvertently trodden on, the stonefish's extremely efficient venom mechanism comes into play. Venom is injected through very sharp, strong spines that adorn this creature's back. These poisonous spines produce one of the most painful and dangerous stings to come from a sea creature. Stonefish are found on reefs and in shallow lagoons along the shorelines of tropical Australia. Their threat to man lies mainly in their sedentary habits—they are quite undisturbed by the approach of larger animals, secure in the knowledge of their lethal defence mechanisms.

Stonefish and other related poisonous fish seem to use their venom mechanisms purely for defence. Oddly enough, all of them are edible when their stinging spines and associated poison glands are removed.

Marine and estuarine stingrays are also capable of inflicting nasty stings. In about two-thirds of stingray species, their barbed spines are venomous—envenomation can cause intense pain but is rarely life-threatening. However, two cases of death from stingrays have been recorded in Australia, both from wounds received in the chest—one victim was actually stabbed in the heart. Rays usually lie on the bottom of the ocean floor, using their barb to deter any approaching potential enemy by lashing their spine sideways or curving it upwards in the direction of the attack. Luckily for beachgoers, stingrays are timid creatures, flapping away at the first signs of approach. They are also active mainly at night.

Pufferfish, bizarre and beautiful, but extremely poisonous to eat.

A number of sea snakes inhabit Australia's tropical waters. Unlike the fish species already described, who use their venom for defence, sea snakes are active stingers, using their venom mechanism for hunting and killing fish. The snake bites the fish, often in mid-body, and grips on to it, still pumping venom, until its victim is immobilised. It then works it around to its mouth and swallows it whole in the same manner as a land snake. These inquisitive reptiles often approach divers but there is little evidence to date of aggressive behaviour. They should, however, be regarded as potentially dangerous as their venom is highly toxic.

Another dangerous marine creature is the cone shell, which is much sought after by collectors for its ornate and decorative patterns. Found on reefs, or crawling along sandy ocean floors, the four hundred or so species of cone shell all possess highly developed venom apparatus. They prey on worms, snails and small fish whom they capture by harpooning with a sharp hollow dart held in the end of their proboscis which extends from the shell. Venom is then injected into the victim, rapidly affecting the neuromuscular system, and paralysing the prey, which is then drawn back into the mouth of the cone. Cone stings on humans take the form of a small puncture, but the effects can be very severe and have been known to cause death.

Octopuses and squids also have venom-producing apparatus, used in conjunction with their horny jaws for capturing prey. The most dangerous species of octopus—and one of the deadliest creatures in Australia—is the tiny blue-ringed octopus. A common sight in rock pools and on reefs almost anywhere on the mainland or Tasmanian coasts, this comical-looking creature can inject a dose of highly toxic venom, capable of rapid diffusion throughout the body. The venom is held in two salivary glands, each as big as the octopus' brain. In a mature specimen there is enough venom to cause the paralysis and eventual death of ten men. Only a small portion of this venom store is delivered in a single bite, but on two known occasions in Australia, it has been sufficient to kill. Luckily, this octopus is not aggressive and bites only when provoked, particularly when taken out of the water. When feeding, the octopus either grasps its prey—usually a crab—with its arms and pierces the shell with its beak, injecting venom, or else it hovers over it and squirts venom into the water, waiting for paralysis to set in before levering the shell open and devouring it.

One extremely dangerous creature found in Australian waters is the sea wasp or box jellyfish. This jellyfish is the world's most dangerous marine stinger and is distinguished from other less dangerous jellyfish by its box-shaped body with long venomous tentacles suspended from four corners. These tentacles contain microscopic barbs of extremely potent venom which, on contact, is instantly injected into its victims. If a swimmer strikes the long streaming tentacles, they cling to him and this action causes the thousands of venom capsules to trigger—death has been known to occur within two to three minutes. In the far north of Australia, sea wasps are blown towards the coast by seasonal winds in the hot, sultry, wet summer months when swimming is most attractive. Hence the likelihood of incidents is great. Sea wasps have translucent bodies and are virtually invisible in the muddy or cloudy inshore waters which they favour. When feeding, they kill their prey, which varies from small crustaceans to fish, by injecting venom from their tentacles. The prey is then drawn up under the body where it is digested.

Another form of marine stinger that is a danger to swimmers on Australian beaches is the bluebottle or Portuguese man o'war. Like sea wasps, the bluebottle subdues its prey with venom from cells along the tentacles which can be 3 or more metres long. It can inflict

Scourge of Australia's swimming beaches, the bluebottle or Portuguese man o'war.

painful stings on swimmers, but no fatalities have been recorded. Blown ashore by summer winds, often in plague proportions, this animal consists of a bright blue gas-filled float from which trail long tentacles, several of which resemble thick string. The float acts as a sail when the wind strikes it and the Portuguese man o'war drifts over the oceans at an angle of 45° to the wind.

Insects and Arachnids

Spiders probably cause the most deep-seated revulsion of all creatures that creep and crawl. They are members of a special class of animals called Arachnids who are characterised by having four pairs of jointed legs, two distinct body segments and jaws that consist of clawed pincers. They lack antennae and wings. There are more than two thousand species of spiders in Australia, found in all environments. All are carnivorous and most are armed with venom to subdue their prey; but only a very few have venom toxic enough to kill or disfigure humans. The Sydney funnelweb and the widely distributed redback are the two lethal species of spider in this country, with over twenty-five known deaths being attributed to them. Recent antivenom production (funnelweb in 1985 and redback in 1956) has greatly reduced the likelihood of fatalities.

While the redback and funnelweb are the only spiders established as being lethal, many others have the potential to give a nasty bite. These include the mouse spider (*Missulena occatoria*), the sac spider (*Miturga gilva*), the black house spider (*Badumna insignis*), some of the huntsmen species (*Olios* genus) and Australia's biggest spiders, the barking or whistling spiders (*Selenocosmia* genus).

Recent research has also revealed the cause of the extremely rare phenomenon now known as necrotising arachnidism, where human tissue is gradually eaten away by spider venom. Bites from the white-tailed spider and the wolf spider are thought to cause this skin destruction.

There are two other arachnids which are dangerous to man—ticks and scorpions. Australia has several scorpions, widely distributed throughout the country. The biggest are found in tropical regions, and these also deliver the nastiest venomous bite.

Although tiny and not nearly as fearsome-looking as scorpions, ticks are far more dangerous. The minute blood-sucking paralysis tick caused over nineteen deaths—mainly children—in the first half of this century.

Australia's insect life is richly diverse, with over 54 000 known species of insect inhabiting every ecological niche. It is therefore hardly surprising that amongst this diversity there should be those that are harmful to man.

One order of insects—the Hymenoptera—contain species that are renowned for causing pain or even death: the honey bees, social wasps and stinging ants. In defence of their communities these insects will sting anything they perceive as a threat. In all societies of Hymenoptera the females who are not destined to become egg-laying 'queens' undergo arrested sexual development, becoming neuters. Their secondary reproductive glands produce venom and they develop stings, taking on the role of guardians of the community as well as fulfilling their other functions of food-gathering, nest building and tending new generations of larvae.

An insect sting, *per se*, is hardly likely to kill, but allergies to these stings are becoming an increasing danger. In most of the population a sting from a wasp, bee or ant causes pain, swelling and itchiness, but an unfortunate minority of people can have life-threatening reactions to such a sting. Bees kill one or two people every year in Australia and many other people suffer severe allergies to ant and native wasp stings.

Crocodiles

Crocodiles, the most ancient and majestic survivors of the reptile epoch well over 100 million years ago, are one of the most fear-inspiring animals in Australia today. There are two species: the ferocious estuarine or saltwater crocodile and the less aggressive freshwater species. Superbly adapted to their environment, crocodiles have thick, leathery skin with a covering of bony plates, providing an armour strong enough to deflect bullets striking at an angle. Their teeth, like those of sharks, can be replaced if lost. Since crocodiles usually swallow their victims whole, these teeth are mainly used for gripping, however, larger kills are torn apart by the teeth with a wrenching force supplied by the rotation of the reptile's body. Strong front feet enable the females of the species to be efficient nest builders, and the heavily webbed back feet are used for in treading water when the animal is floating. The powerful tail is used to stun prey, as well as for propulsion and steering when the crocodile is swimming. Eyes and nostrils are so placed that the crocodile can swim or float with

The Blue Mountains funnelweb (*Hadronyche versuta*).

The saltwater crocodile—an object of fear and repulsion.

only these showing and the nostrils can be closed when the animal is submerged. Feeding patterns vary, but large victims are usually seized by the head and then turned over and over in the water until they drown and can be eaten.

No-one really knows how long crocodiles can live or how large they can grow. Like all reptiles, they keep growing throughout their lifetime, becoming more skilful and powerful as hunters. And, apart from man, they have no predators. Now legally protected, crocodiles reign supreme in the waters of northern Australia. It is up to us to treat them with respect and take no risks in crocodile infested waters.

The Cane Toad

The only toad found in Australia, the imported cane toad ranges over the north-eastern part of the continent. Like all toads, this species has enlarged toxin-producing parotid glands which appear as prominent lumps on each side of the head behind the eyes. When squeezed, they produce poisonous fluid which can be squirted to a distance of one metre. This venom is highly poisonous to small animals, and larger animals are killed within a few minutes of biting a toad. If this toxin comes into contact with humans it can cause accelerated heartbeat and breathlessness, and to eat the poison-riddled flesh of a cane toad would cause devastating illness.

Toads shelter under logs and debris during droughts, breeding when temperatures reach about 26°C. Breeding takes place in any temporary or still waters. Female toads lay about 35 000 eggs a year; and, to further ensure continuation of the species, when females are scarce males develop ovaries and turn into females.

Originally introduced to the Queensland cane fields in an unsuccessful bid to control the cane beetles, these toads have now become an obnoxious pest themselves, spreading up the Cape York peninsula, into far western Queensland to beyond Mt Isa and into the Northern Territory.

Ground dwelling, they are usually found at night foraging on the forest floor or on roadways. They have a warty skin covering and are grey, olive-brown or reddish-brown above.

Land Snakes

About 60 per cent of all land snakes found in Australia are poisonous—Australia has a higher proportion of venomous snakes than any other country in the world. Fortunately, most of these are small and of little danger to humans. Only about twenty species pose a serious threat to man, and all of these belong to the one family, the Elapidae. This family contains some of the most dangerous snakes in the world, including the African mambas, African and Asian cobras and the Asian kraits. In Australia, the most dangerous elapid snakes are the taipan, the death adder, the tiger snake, the copperhead and the brown snake. All can be found in quite highly populated areas in the country and in newly developed suburbs.

Snakes represent one of the great successes of specialised development. Descended from the same stock as lizards, these reptiles have evolved over the years to claim their own ecological niche in widely varied habitats. Their graceful, sinuous movement is derived from their bone structure, based on a backbone containing up to four hundred vertebrae. Each vertebra is connected with a movable rib, through a ball-and-socket joint, and muscle chains run in all directions. The result of this internal structure is the typical flowing action of the snake's progression, in which the snake's whole body follows the same winding course. Muscular waves pass along the snake's body, fractionally lifting successive sections of the body and pushing it forward, while the belly scales catch on any irregularities in the ground.

The most highly developed sense in snakes is that of smell. They have a method of 'tasting' the air by using their long forked tongues. Constantly flicking in and out, the tongue picks up particles in the air and transfers them back to sensory organs in the roof of the mouth. The snake's vision is only acute at short range and their hearing is inefficient as they have no ears and can only hear through an internal arrangement which allows them to pick up little more than ground vibrations. All land snakes can climb trees as well as swim and they can lift the front third of their body clear of the ground—as when they attack.

Most dangerous snake species are viviparous, giving birth to live young. Exceptions include the taipan and brown snake who are oviparous, laying leathery eggs that take weeks to incubate. Young snakes are born or

A lethal inhabitant of sandy scrubland areas, the death adder basks on a log awaiting unwary victims.

King of Australian snakes, the taipan is the biggest, deadliest, fastest and most intelligent of all.

hatched fully formed and receive no attention whatever from the female snake. Within days they are able to kill tiny prey. The lifespan of a snake is thought to be from ten to thirty years.

All snakes are carnivorous, their food ranging from small insects to entire wallabies. Their fine, sharp, backward-pointing teeth are designed for gripping, not chewing, prey, so everything must be swallowed whole. To assist in this, the snake's jaws come disjointed at the back and other bone connections enable the mouth to stretch enormously. Powerful, fast-acting digestive juices enable snakes to digest whole bodies as well as lubricate the passage of the victim through the alimentary system. Like all reptiles, snakes are cold-blooded and are only fully active at temperatures between 20° and 30°C. In any other climates they become sluggish and if the temperatures are extreme, die.

The snake's venom glands, evolved from its salivary glands, are situated in the temple, just under the skin. When a snake bites, its venom is expressed from these glands and is squeezed along a duct to the base of each fang. The two curved fangs have a groove running down from the root to the point and the venom travels down this into the tissue or bloodstream of the bitten animal, rapidly immobilising it. Fangs are subject to wear and tear and often work loose. They are, however, soon replaced as most snakes have three or four reserve fangs lying in the jawbone behind the functional pair, ready to grow forward when required.

Snakes do not always inject venom when they bite. They may choose to launch a mock strike with their mouths closed, or even open their mouths and strike without injecting any venom. The amount of venom injected can vary, too, depending on the snake's previous activities. The output is slight if he has made a recent kill, but copious if he has gone hungry for long.

Snake venom is a powerful multi-purpose enzymic fluid which contains four main constituents. The characteristics of each type of snake venom are different and the effects on the victim depend on the predominant toxin present. The most potent constituents are the neurotoxins which block communication between the nerve ends and muscles, bringing about general paralysis of the victim. An enzyme called haemorrhagin breaks down tissue, particularly blood vessels and internal organs. Haemolysin destroys red blood cells and thrombase causes rapid coagulation of blood and the blocking of blood vessels.

The primary purpose of this venom is the rapid subjection of small prey for consumption. Snakes are not normally aggressive towards man. They are generally secretive and shy and will glide away at the approach of man if given the opportunity. Most snake bites and attacks stem from accidental contact such as treading on one, or occur when someone is attempting to kill or catch the snake. Snakebite statistics are not well-documented.

Antivenom usage probably gives the best indication of the occurrence of serious snakebite and the Commonwealth Serum Laboratories report about two hundred cases annually. Typical symptoms of snakebite are drowsiness, severe headache, nausea, vomiting, confusion and a fast pulse. Double vision, difficulty in swallowing and loss of consciousness can follow.

Antivenom is the only substance capable of neutralising the various components of snake venom, and there are now antivenoms for most species of dangerous snake. Tiger snake antivenom was produced in 1930; taipan in 1955; brown snake in 1956; death adder in 1958 and black snake in 1959. A polyvalent antivenom, which combines all the land snake venom groups, was released in 1962 to combat poisonings by unidentified snakes. Antivenom is simply serum from an animal immunised against venom. Fortunately, snake venoms are antigenic and therefore capable of creating specific antibodies in a mammal's bloodstream. Most antivenoms derive from horse blood. Venom is obtained by snake-milking—allowing the snake to bite through a rubber diaphragm stretched tightly over the mouth of a collecting vessel. The venom discharged into the vessel is subsequently dried and then injected in solution in small but increasing doses into a horse. As the horse is a large animal, the venom has little effect, other than to stimulate the production of antibodies in the blood. When the concentration of these antibodies is high enough, blood is taken from the immunised horse and treated chemically to yield a substance which can be injected into human snakebite victims to destroy the venom.

Despite the prevalence of dangerous species of snake in Australia, fatalities from snakebite are now rare. However, each State has its share of these venomous reptiles and we should leave them well alone.

The woma (*Aspidites ramsayi*), a terrestrial python, seen here near Ayers Rock.

The comical toadfish, pufferfish, or porcupinefish (*Dichotylichthys* sp.) contains a lethal toxin, less than a gram of which is enough to kill anyone. The poison acts faster than a snake bite. About thirty species of pufferfish occur in Australian waters and they are one of the easiest fish to catch with a handline and hook. The fish are safe to handle but should not be eaten. The poison, tetrodotoxin (TTX), is a hypotensive agent which causes neuromuscular weakness, producing neuromuscular block, hypotension and respiratory depression. The fish's liver, ovaries and eggs contain the heaviest concentrations of TTX. In Japan the fish is considered a delicacy and specially licensed chefs, after much involved preparation, present certain parts for consumption, under the name 'fugu'. Pufferfish often inflate themselves when handled, or as a defensive ploy, by gulping in water. The water is held in the stomach until released. If left out of the water they gulp in air in the same manner.

The sharpnose pufferfish or black saddled toby (*Canthigaster valentini*). Pufferfish can range from 5–75 cm in length. They are slow weak swimmers and are usually seen alone. There are two families of this strange-looking fish in Australian waters: the Tetraodontidae and the Diodontidae. All have unique teeth which are fused together. The Tetraodontidae have a median cleft to the fused teeth, which gives the appearance of four beak-like teeth. The Diodontidae lack the median cleft. Their skin does not have the scales of the bony fishes and may be quite smooth or covered with prickles. Colour is diverse and shape is mostly ovoid or slightly angular. One of the larger species of pufferfish, the silver-cheeked toadfish (*Gastrophysus scleratus*) is said to have caused the near-death of Captain Cook when he sampled some of its cooked flesh in New Caledonia in 1774.

The deadliest octopus in the world—the blue-ringed octopus. This dainty pretty mollusc is extremely plentiful: there are two species, *Hapalochlaena lunulata* which grows to 20 cm and is common in the north, and *H. maculosa* (*above*) which grows to 12 cm and is found around southern shores, particularly in summer. Apart from the difference in size, the species are virtually indistinguishable. Both have been known to kill humans. When provoked they will make an incision into the skin with their parrot-like beaks, forcing venomous saliva into the wound. Paralysis sets in, within minutes of the bite in severe cases, and death from respiratory failure can follow. The venom contains tetrodotoxin (TTX) identical to the poison found in the tissues of pufferfish. Luckily, this octopus is not aggressive to humans and rarely bites unless taken out of the water.

At rest in a rock pool the blue-ringed octopus is rather inconspicuous, but when disturbed, and particularly when taken out of the water, its colour pattern darkens dramatically, the dark rings becoming iridescent peacock blue. Some beachcombers are attracted by its beauty and vivid colour changes—but should beware of picking it up! A shell-less mollusc, the octopus' brain, eyes and mouth are clustered over the junction of its eight arms which are adorned with harmless sucker pads. The stomach and other body organs are carried in the bulbous upper part. The blue-ringed octopus differs in several ways from other species of octopus. Its ink sac functions only in infancy (most octopuses squirt ink to deter predators); it lays only 100–200 eggs which stay fastened to the mother's arms until they hatch (other species lay thousands of eggs and attach them to rocks or plants); and they mate during a close embrace (in other octopuses the male uses the tip of one arm to transfer sperm from his body cavity into that of the female). Blue-ringed octopuses breed only once and the females die shortly after hatching their eggs. Their life span is about seven months.

An evil inhabitant of coral reefs and estuaries, the stonefish is the most venomous fish in the world. There are at least two species of this nasty creature: the reef stonefish (*Synanceia verrucosa*) and the estuarine species (*S. trachynis*), pictured. Its venom can cause rapid and agonising death. Sedentary in its habits, the stonefish is undisturbed by the approach of larger animals and anyone wading through shallow pools is in danger of stepping on one without seeing it. Along the back of the fish is a row of thirteen extremely sharp, strong spikes which have large venom glands. Venom is released when pressure is exerted on the glands—as may happen if a person steps on the fish and the spines penetrate the sole of the foot. Severe pain, delirium, swelling, temporary paralysis and symptoms of shock may develop. Many Pacific and Indian Ocean islanders have been killed by stonefish, but no deaths have been recorded in Australia.

The singularly ugly stonefish has skin covered with warty protuberances, often coated in algae and marine encrustations. A master of disguise, this fish is virtually undetectable when lying on dead coral or rubble on the sea bed. It often buries itself in the sand or mud, lying in wait for smaller fish to pass, which it then engulfs with its enormous mouth, sucking them in with lightning speed. An ungainly swimmer, the stonefish is heavy bodied and stumpy tailed, with huge side fins used for scooping out prey from hiding places. However swimming skills are little needed by this creature, whose greatest predatory successes arise from playing a waiting game. It lives on reefs and in shallow lagoons of tropical Australia and has been found in scattered locations between the Houtman and Abrolhos Islands, off Geraldton, Western Australia, and off far northern New South Wales as well as along the Great Barrier Reef.

The Portuguese man o'war or bluebottle (*Physalia*) is probably the best known of all Australia's venomous marine animals. The long trailing tentacles, well-armed with stinging capsules, can cause intense irritation and a rash which can last from several hours to a few weeks. This creature has a bad reputation on summer beaches when inshore winds bring in swarms of the stingers, driving swimmers from the water in their hundreds. The bluebottle is not a jellyfish, but one of a related group of coelenterates, the siphonophores. It is not one individual, but comprises a colony of individual organisms, born from one egg but developed and specialised to perform different functions. One organism inflates itself with gas and creates a float, others make up the tentacles and still others form the feeding polyps and the reproductive parts. All work together for the survival of the group organisation.

(Below and right)
The sea wasp or box jellyfish (*Chironex fleckeri*) is the most potentially dangerous of all marine stingers. It has caused the death of about seventy people—mostly children—in Australia's tropical inshore waters since the late 1800s, a figure exceeding the combined toll taken in the same area by crocodiles and sharks. Death from the sea wasp's sting has occurred within thirty seconds. The danger period for these creatures is during the summer months in Australia's tropical northern waters. They prefer the cloudy inshore waters and have not been found in the clear waters of offshore islands and coral reefs. A very active jellyfish, the sea wasp can propel itself through the water quite rapidly, in the manner of an octopus. It is the most advanced jellyfish known, with a well-developed nervous system. It reacts to water disturbances, and sensors around its bell can detect large objects by changes in light intensity.

The semi-translucent bell of a box jellyfish or sea wasp is extremely difficult to see in the water; a purplish tinge near the tentacle base may be all that is visible. The cube- or box-shaped bell can measure up to 20 cm across and 30 cm deep. Tentacles are sprouted from four appendages protruding from the open end of the bell. A big

specimen may have up to fifteen tentacles in each bunch, giving it a total tentacle length of nearly 100 metres. Thousands of tiny stinging capsules are distributed along this jellyfish's tentacles. The stinging capsule is unique, containing a thin tube, coiled and compressed like a watch spring. When stimulated, the capsule breaks open and the tube is forcibly shot out to penetrate the skin of its prey. The capsule also contains an extremely potent venom which is injected into the victim through the hollow tube. The dose of venom depends on the length of the tentacle that comes into contact with the victim—adhesion of 6–7 metres of tentacle means probable death to untreated victims. Early symptoms of a sea wasp sting include immediate extreme pain and whitish stings stuck to the skin which cause red, purple or brown weals. There is an antivenom available.

The leopard moray eel (*Gymnothorax flavimarginatus*) is a reef-dwelling species found in tropical waters. These savage creatures tend to hide in crevices and under rocks during the day, with only their heads protruding to occasionally snap at passing fish. At night they emerge to forage for food. Devoted to their lairs, they ferociously defend them against intruders and can inflict a serious wound if threatened. Although not venomous, a moray eel bite on a human victim can be deep and ragged and prone to serious infection. Morays can exceed 3 metres in length, with bodies as thick as a man's thigh. The world record length for an eel is claimed for a 3.7 metre moray taken in the Coomera River in Queensland.

Other species of moray include the green moray (*Gymnothorax prasinus*), common on rocky reefs off all temperate coasts, and the mosaic moray (*Enchelycore ramosa*) found in rocky reefs off New South Wales and the coral surrounding Lord Howe Island.

(*Left*)
Moray eels have extremely powerful jaws and long fang-like depressible teeth, ideal for seizing prey that is too large to be swallowed whole. Marine eels are carnivorous, feeding on smaller marine animals including fish, crustaceans and squid. Smaller prey are swallowed whole, while chunks are torn from larger prey. To do this, the moray locks its jaws onto the fish then ties itself in a knot, powerfully contorting its body to wrench away a huge piece of the fish's flesh.

Sharks have the most powerful jaws in the world—the jaw power of a 2.4 metre shark, measured on a bite meter, has shown a pressure of 277 megapascals. Their razor sharp teeth are not set into the jawbone and are often lost. To combat this, all sharks have extra teeth in their jaws and whenever one is lost a replacement tooth swivels up from inside the jaw tissue. A shark can use up more than 20 000 teeth in ten years. Most sharks have rows of teeth, one behind the other. Except for the front teeth they lie flat like shingles on a roof. Tooth shapes vary considerably between species, for example a blue pointer has pointed fangs and a tiger shark has broad teeth with a saw-edged shoulder. Pictured is the jaw of a white pointer shark.

Sharks have been blamed for killing about 150 people in Australian waters following the 1790s. Ocean bathing first became popular on the east coast beaches between the world wars and during this period fatalities averaged three a year. In the twenty years following 1937, when shark meshing off popular Sydney beaches was first undertaken, fatal attacks fell to less than two a year. Most of these attacks occurred in north Queensland. Since 1957, the national rate of fatalities has dropped to fewer than one a year. Shark meshing and spotting patrols by aircraft and helicopters have virtually eliminated shark attacks from many surf beaches on the east coast of the country. Nets up to 300 metres long are stretched parallel to the beach, about 400 metres offshore. The nets are 6 metres in depth and are floated and weighted so that they touch the bottom. By no means a complete screening, the nets are only set at each beach four times a month, for periods as brief as twenty-four hours. They do, however, cause the decline of local shark populations, a fact proven by the dwindling numbers that are found trapped in the nets.

Tiger sharks (*Galeocerdo cuvieri*) are the terror killers of the Pacific, probably attacking more Australians than any other shark. This species is widespread in tropical and subtropical oceans and occurs all around the coast of Australia, frequently entering bays and estuaries in the summer. The largest Australian specimens were caught off Newcastle, New South Wales, in 1954 and off Mackay, Queensland, in 1980. Both weighed more than 1500 kg and were 5.5 metres long. Most tiger sharks, however, are about 3 metres long. These sharks have thick powerful bodies, short pointed snouts and long finely tapered tails. Juveniles have dark back stripes which fade with age; adults are greyish brown above and cream below. Their teeth are quite distinctive, being broad, notched, serrated and oblique, each shaped like a rooster's comb. Generally a fairly sluggish creature, the tiger shark is a confirmed scavenger and can often be found in harbours and near ships looking for discarded food. When aroused it is a powerful and aggressive attacker.

(Below)
The most formidable of all sharks, the white pointer or great white (*Carcharodon carcharias*). This ocean-ranging species is common in the cold waters of the Great Australian Bight. Its mouth holds large jagged triangular-shaped teeth, able to cut through bones with one bite. Omnivorous, the white pointer feeds on fish, sea turtles, seals and even sea birds, swallowing anything up to half its size with one gulp. It is an extremely active hunter, often raising its head out of the water to look for prey on the surface. This fearsome creature can be distinguished from other sharks by its relatively small conical head which comes to a sharp point in front. Growing to over 7 metres in length, the white pointer is predominantly grey with a whitish underside. Although never found in great abundance it is responsible for over one-third of human fatalities where the attacking species has been identified.

Several species of hammerhead shark are found in Australian inshore waters. The great hammerhead (*Sphyrna mokarran*) and the scalloped hammerhead (*Sphyrna lewini*) pictured, are mainly tropical but can be found in northern New South Wales. The smooth hammerhead (*Sphyrna zygaena*) is common in cooler southern waters. The head of these strange-looking sharks is shaped like a long pole and is at right angles to its body. Its eyes bulge at both ends of the pole and various other sensory organs are distributed along it. A hammerhead can grow to 5 metres long and its head can be a metre wide. They usually hunt in packs and are presumed to be dangerous although there is little evidence of attacks on man.

The beautiful scorpionfish, like their stonefish cousins, are armed with hollow spines in the dorsal and pectoral fins which act most efficiently as a hypodermic syringe for the injection of venom, inflicting maddening pain and symptoms of shock. More mobile than stonefish, they are less likely to be trodden on, but extremely likely to turn up in a fishing catch, posing a particular menace to fishermen. They inhabit estuarine and nearshore waters all around the coast of Australia. One of the most spectacular sights in the marine world, these fish have bony, ridged heads, decorated with zebra-like stripes and carrying strange, tassel-like appendages with waving filaments over the eyes. Their dazzling assortment of feathery, slowly waving fins are marked with vivid stripes of cream, yellow, pink, brown, or scarlet. This family includes the lionfish or butterfly cod (*Pterois volitans, overleaf*) and the butterfly scorpionfish (*Dendrochirus zebra, above*). Both these species are found on the mainland and offshore reefs in tropical and subtropical waters.

Attractive but venomous, the cone shell (*Conus* sp.) is mostly found in Australia's tropical waters—particularly near the Great Barrier Reef. Cones manufacture hollow barbed darts, as long as 1 cm, which project from their snouts and can extend to the length of the shell. The cone rams its snout against its victim—marine worm, mollusc, or fish—and the dart, flooded with venom, is driven in. The paralysed victim is then drawn back into the snout which can widen enough to consume prey nearly as large as the cone itself. One species, the geographer cone (*Conus geographus*), grows to 13 cm and is a known killer of humans overseas and the only proven killer in Australia. Initial symptoms of cone envenomation are usually immediate, with intense pain followed by respiratory failure and general muscular paralysis. Death, if it occurs, probably results from heart failure. Little is known about which species are safe to handle and which are highly venomous, but all should be treated warily as, at the least, they can inflict intense local pain.

Conus species known to be lethal include *Conus striatus* (pictured) which grows to 10 cm, and *Conus tulipa* which grows to 7.5 cm. Active hunters at night, by day cones rest among coral rubble or under stones, often burying themselves with only their siphon visible. The siphon draws in water and pumps it over the cone's breathing gill.

The black stingray (*Dasyatis thetidis*) frequents Australia's temperate waters. Stingrays possess a barbed spine projecting from the base of the tail; the edges of this spine carry rows of sharp backward-pointed barbs which can tear and lacerate flesh. Two grooves, containing venom glands, run along the top of the spine and a venomous secretion can enter the gashes made by the barbs, causing intense pain in the victim. Stingray venoms, however, are not directly life-threatening; of more concern is the high likelihood of infection in the wound. Most injuries occur to people wading in shallow waters or to fishermen handling accidentally netted rays.

Stingrays commonly lie on the ocean floor, or bury themselves in the sand or mud. They are related to sharks, but instead of following the aggressive, predatory evolutionary path of their cousins, they evolved other mechanisms for defence and the capture of prey. Their barbed spine is used to deter potential enemies and instead of having the mighty cutting teeth of sharks, they have small grinding teeth to feed on their diet of molluscs and crustaceans.

The yellow-bellied sea snake (*Pelamis platurus*) often enters inshore water and is sometimes beached in storms. Most sea snakes in Australian waters are found in the tropics, sometimes riding the warm currents south in summer. The yellow-bellied, however, is found in the south throughout the year. Over thirty of the world's fifty-odd sea snake species have been identified in Australian coastal waters. Although the venoms of most sea snakes are highly toxic and fast-acting, even the biggest species have relatively small venom glands, low outputs and short fangs rendering them not as lethal as the land snakes. There is little evidence that they are aggressive. They seem to bite only when held or squeezed. There is no record of an Australian fatality from a sea snake bite.

Like land snakes, sea snakes are air breathers, coming to the surface at regular intervals to breathe. Adaptations to marine life include a flattened oar-shaped tail—effectively displayed by the olive sea snake (*Aipysurus laevis, overleaf*)—upward-pointing nostrils with flaps that close when the snake submerges, and an enlarged lung that terminates in an air sac which acts as a reservoir for air when diving.

(*Below*)
The black-banded sea krait (*Laticauda colubrina*) differs from a true sea snake in that it lays eggs, going ashore to do so. True sea snakes bear their young live, at sea. Sea kraits have the same wide belly scales as land snakes and adapt their breathing rhythm to suit water or fresh air. The black-banded krait averages about 1 metre in length and the specimens occasionally found in Australian waters are usually strays from western Pacific breeding areas. Although usually inoffensive, their venom is potent and they are dangerous if provoked.

Found across the whole continent of Australia, the deadly redback spider (*Latrodectus hasselti*) is the greatest single cause of serious envenomation in this country, far outnumbering snake bites and marine stingings. The prominently marked female redback has well-developed venom glands that inject enough venom to kill a man. The symptoms of a redback bite begin with a stinging sensation, with the bitten area usually reddening, swelling and often sweating. General symptoms of severe pain, nausea, vomiting, sweating, muscle weakness, accelerated heart rate and raised blood pressure follow. Luckily, the neurotoxic component of the venom is slow-acting and there is an antivenom available which can successfully be administered some time after a bite. About 300 cases per year of redback bites have required antivenom in recent times.

Redback spiders inhabit dry, dark places—in logs, under bark, in foliage, under roof-eaves, in rafters, in outbuildings and under refuse in rubbish dumps. Their webs are of a tough bluish silk, tethered by guy lines to nearby objects. The female redback is relatively small, generally no longer than 12 mm. Her body is dark brown to black and covered with short hairs, giving it a velvety sheen. There is a distinctive hour-glass design under the large globular abdomen, and usually a red longitudinal stripe on the back. The harmless male redback is smaller (3 mm) and has more sombre markings. The female spins several egg sacs (*below*), laying up to 300 eggs in each—not all of these hatch as they are heavily preyed upon by wasp larvae.

The most dreaded creature of the Sydney area—the Sydney funnelweb spider (*Atrax robustus*)—is probably the world's deadliest spider. Since 1927 this spider has killed at least thirteen people. Bites by both sexes of funnelweb are immediately painful but the male is the killer. Symptoms of general poisoning can show within ten minutes, with severe pain, general sweating, numbness around the mouth, nausea and vomiting, delirium and respiratory failure. Death can occur within a few hours, particularly in children—one two-year-old child died within fifteen minutes of a bite. An antivenom has been available since the early 1980s, saving many critically ill bite victims.

Funnelwebs are heavily built, large and black with shiny smooth legs. Their fangs, up to 7 mm long, can pierce the skulls of smaller prey. When irritated, the spider will rear up to assume a striking position, globules of poison clearly visible at the tips of its downward-striking fangs.

A highly distinctive feature of the funnelweb's burrow is a system of surface threads stretching out from the entrance hole. These are to warn the spider when prey is within striking distance. Waiting just inside the burrow, the funnelweb darts out when its web is disturbed. Its usual victims include insects, slaters and snails. The silk-lined burrows with funnel entrances are always made on the ground in cool, humid, sheltered spots such as rock crevices, garden shrubberies or among debris. Breeding time is the danger time. Mating activity occurs through summer and autumn, and during this period the male funnelwebs roam widely, searching for a mate.

Redbacks and funnelwebs have always been considered the only really dangerous Australian spiders, but recent research has uncovered a new threat. An unknown ingredient of some spiders' venoms eats away tissue, in severe cases eroding through all the skin layers. Known as necrotising arachnidism, this phenomenon, relatively uncommon, has only been documented in Australia since the 1970s. The tissue destruction can be extremely severe, requiring skin grafts and in some cases amputation. It has not been scientifically proven which spiders are responsible for this tissue death, but the two major offenders appear to be the white-tailed spider (*Lampona cylindrata, above*), a 1.5 cm long creature that is often found in homes, and the wolf spider (*Lycosa godeffroyi, below*), a 3 cm long aggressive ground-dwelling species, pictured with spiderlings clustered on her back.

Scorpions have long been regarded as dangerous, but reports of bites in Australia have been few and there have only been two reports of deaths attributed to them—both of babies. Bites can cause pain and swelling but further symptoms of poisoning are rare. Scorpions are generally furtive, sedentary creatures, hiding under logs, stones, bark, and leaf litter. They are found in all habitats throughout Australia, and in all climatic conditions. One of the oldest existing types of animals, they are grouped with spiders, ticks and mites in the class Arachnida. They are easily recognisable by their segmented body, well-developed claws and slender curved tail which bears the sting. The venom contained in the sting is used to immobilise their prey—usually spiders and other insects.

The underside of a scorpion (*left*). Scorpions differ from their spider cousins in that they have no waist between thorax and abdomen and the abdomen extends into a narrow flexible tail, the telson. The telson bears the sting which is fed with venom by ducts from the two venom glands. The scorpion's tail can be thrown over its back as well as swung sideways. The frontal appendages are enlarged into huge, pincer-armed claws.

Bulldog ants (genus *Myrmecia*) are distinguished by their massive size and by their prominent huge-toothed mandibles. Found throughout Australia, they can inflict a sting more painful and prolonged than that of a bee or a wasp. Bulldog ants are ferocious defenders of their nests, rushing out to attack any intruder. Nesting mounds vary in size from low humps to large cones up to 45 cm high, and they are usually constructed on well-drained high ground or slopes. These massive ants—up to 2.5 cm long—kill their victim by first grasping it with their fearsome-looking mandibles and then curling their bodies around to use the formidable venomous sting at the tip of the abdomen. The most primitive of the world's ants, bulldog ants, apart from one species found in New Caledonia, are peculiar to Australia. Elsewhere there are only fossil remains.

Green tree ants (genus *Oecophylla*) are aggressive defenders of their nests and nesting trees, and will drop on to human visitors and sting savagely. Widespread over northern Australia, they colonise mangroves, paperbarks and other trees growing in moist places. Most colonies are formed in the wet season, but the vast majority only last for six months with just a few developing into mature colonies. Green tree ants are weavers, fastening leaves together to form a nest high above the ground, creating a dry shelter for themselves. Nests can be as large as a football.

Wasps of the native paper-nest species, *Polistes,* tend their bell-shaped nest. Paper-nest wasps are widespread throughout Australia, fixing their nests to rock overhangs, tree branches and sometimes the roof-eaves of houses. To make the nest, the queen wasp tears away shreds of weathered wood which she masticates into a grey paste with her salivary juices. This material closely resembles paper. The nests are fixed by a narrow stem around which a number of open downward-facing cells are constructed. There are thousands of species of wasps in Australia, both native and introduced. They can inflict agonising stings, but fatalities are rare.

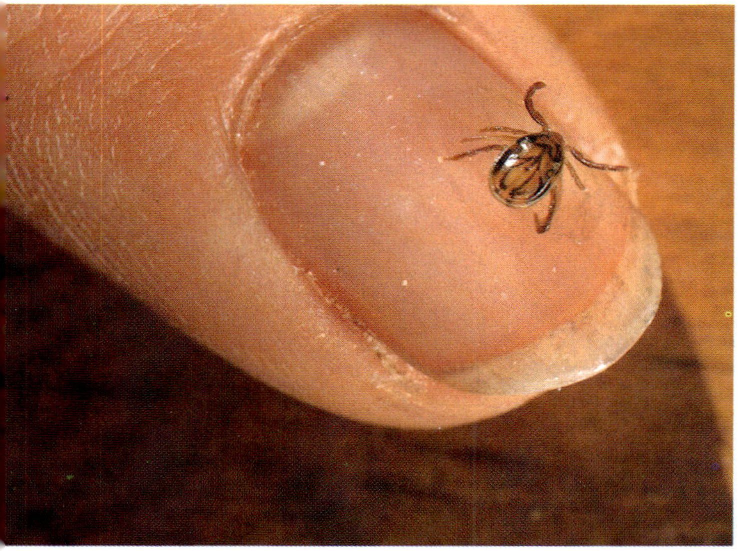

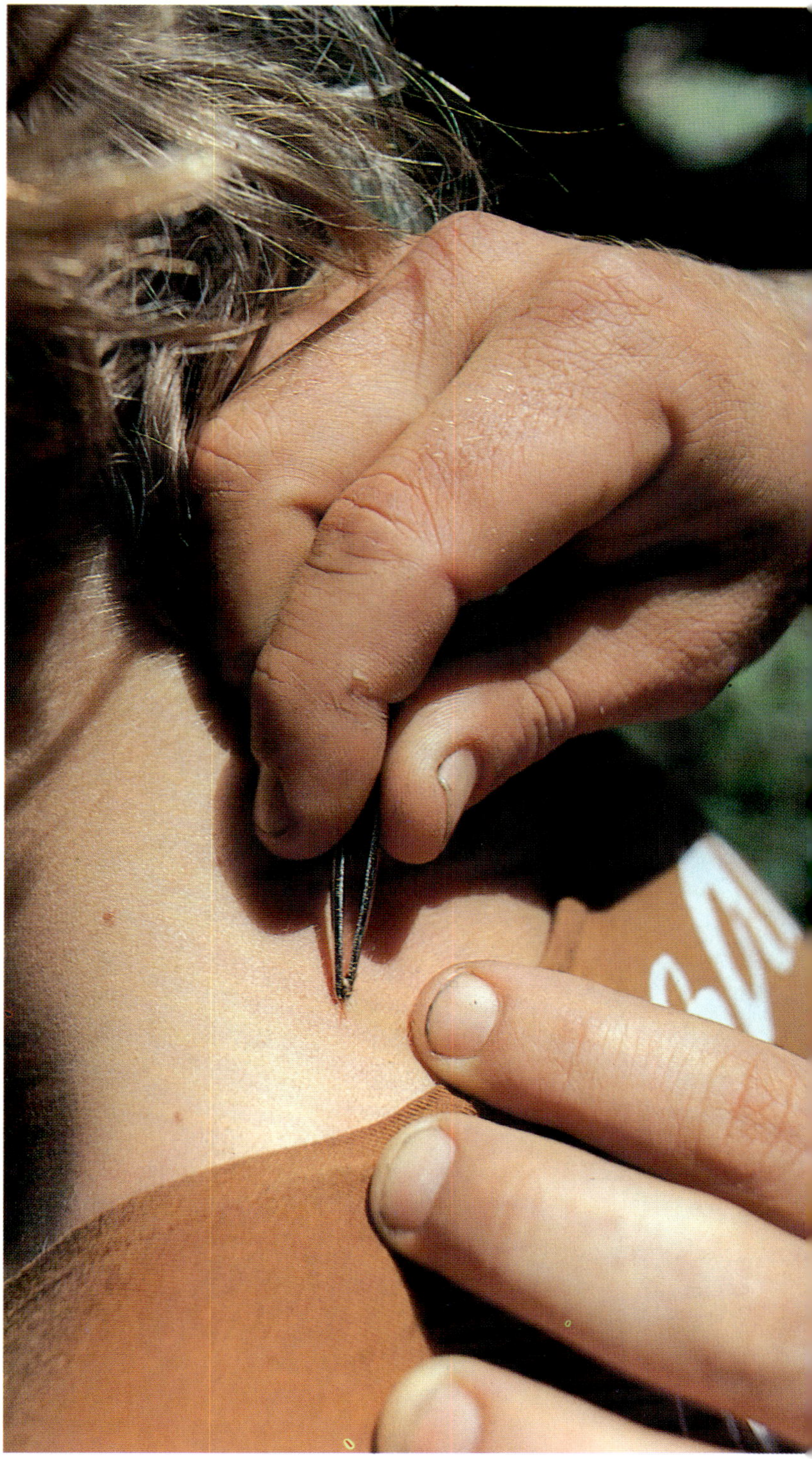

Bites from the adult female paralysis tick (*Ixodes holocyclus*) have ranked as a major killer on the eastern coastal belt of Australia. Of nineteen deaths in New South Wales between 1902 and 1947 seventeen were of infants between one and three years old. Lately, due to scientific advances, death from tick bite seems to be a thing of the past. Doctors are now more aware that paralysis can be caused by the tick and hospitals can usually keep the victims alive until the effects of the toxin wear off.

Paralysis ticks live in bushland and often lodge themselves on an unsuspecting person. The tick grasps the skin with its claws and inserts a barbed tube into its host, alternately drawing on its victim's blood and injecting a fluid from its salivary glands. This fluid is a potent toxin to the central nervous system and can cause severe paralysis.

When first lodged in the host, the adult tick is oval, flat and yellowish, and about the size of a match-head *(left below)*. After it has fed for a while it darkens in colour and swells to the size of a pea. The tick's sole diet is the blood of warm-blooded animals, including native mammals and birds, domestic animals and man. Tweezers should be used to remove a tick as soon as it is discovered. Slip them under the tick's head, around the feeding tube *(above)*, then firmly pull the tick off.

The most sinisterly named of all Australia's snakes, the death adder (*Acanthophis antarcticus*), is one of the most dangerous snakes in the world. This placid snake often lies camouflaged among leaf litter to wait for its prey. If trodden on it strikes with lightning speed and unerring aim. Members of the viper family, death adders have short thick bodies with large triangular heads on a narrow neck. Their tails taper to worm-like projections which they twitch to attract small rodents, lizards and birds to within striking distance. They average 65 cm in length and are nocturnal, frequenting sandy scrubland over most of the Australian mainland. They are variable in colour, ranging from pale grey to a reddish brown, usually with irregular cross-bands. Young are born live, with up to twenty in a litter.

(*Right*)
Great care is needed to obtain snake venom for antivenom production. This death adder is being forced to bite through a rubber membrane stretched over a container. The snake's venom glands are massaged to obtain the maximum amount of venom.

The death adder has developed large venom glands and long fangs—more than 6mm in length. Its bite is usually lethal.

The eastern or mainland tiger snake (*Notechis scutatus, above*) is responsible for a significant proportion of snakebite deaths in Australia. It is not a particularly aggressive snake but is a prolific breeder and ranges over the heavily populated areas of coastal south-eastern New South Wales, Victoria, and south-eastern South Australia; it is therefore probably encountered more frequently than other snakes.

Tiger snakes are normally inoffensive unless aroused, when they will flatten their necks in a cobra-like fashion and hiss before striking. Their venom is one of the most potent in the world. Tiger snakes are largely diurnal but are nocturnal in warm weather and prefer swampy ground to places where there are low shrubs. They average 1.2 metres and usually have cross-bands of yellow or cream on their grey, brown or green backs. Young are born live in an average litter of thirty and are exact replicas of the adult in colour and markings.

The black tiger snake (*Notechis ater, below*) has its main population in Tasmania, and subspecies are found in the Flinders Ranges, King Island, the Yorke and Eyre peninsulas, Kangaroo Island, and south-western Western Australia. In this species the colour banding is absent or barely noticeable.

The tiger snake's diet comprises mainly frogs or tadpoles, hence their distribution over areas with average rainfall greater than 500 mm. When frogs are scarce, they will feed on small rodents, lizards, birds, eels and fish. After striking their prey, they hang on to it with their fine, sharp teeth until it succumbs to the venom and then they swallow it whole. To assist in this, the lower jaw becomes disjointed via an elastic ligament, enabling the snake to swallow a creature larger than its head. The curved teeth on each side of the snake's jaw appear to creep forward alternately as the prey is forced down the neck. To prevent suffocation while eating, the snake breathes though a wind-pipe at the side of its mouth.

The taipan (*Oxyuranus scutellatus*) is the world's deadliest snake. It is the largest of Australia's snakes (averaging 2.5 metres but can grow to over 3 metres) and the fourth largest of the world's poisonous snakes after the king cobra, black mamba and bushmaster. The taipan also has the largest fangs of Australia's snakes—they can grow to more than 12 mm. The venom is squeezed from a pair of glands in the head and runs down the hollow fangs into the wound, which is very deep because of the length of the fangs. If cornered, the taipan will attack ferociously and may bite several times in succession, not bothering to hold on.

This deadly creature is found in well-grassed tropical woodlands in the wetter coastal areas of Queensland, the Northern Territory and Western Australia. The Queensland canefields are a favoured habitat, the rats which inhabit the area being an ideal food source. Taipans have long narrow heads and slim bodies. Their back is brown, belly cream to yellow, and their head is a pale creamy colour. Efficient, alert hunters, they are mainly active by day or at dusk, feeding on rats, small birds, mice, bandicoots and lizards. In very hot weather they may hunt at night.

Taipans are oviparous, laying clutches of ten to twenty eggs which are deposited under logs, rocks or debris. The babies hatch after about three months and are from 35–40 cm in length. These vulnerable young snakes are easy prey for other animals and a large number do not survive.

The mulga snake (*Pseudechis australis*), used to be known as the king brown, but is no relation to the brown snakes (genus *Pseudonaja*). They are similar in general colouration and prefer the same dry habitat, but the mulga can be distinguished by its heavier body and broader head which is slightly distinct from its neck. This snake is extremely dangerous as it has the greatest maximum venom output of all Australian snakes. It holds on to anything it bites, chewing vigorously while pumping venom into the victim. The venom causes paralysis, acting on the muscles rather than directly on the nervous system as is the case with most other species of snake. The mulga is widely distributed throughout Australia, found in many habitats from desert to open forest, and is especially common in semi-arid territory. They are active by day in the south and by night in the tropics, feeding on rodents, lizards and birds as well as other snakes. Along with the death adder, taipan, tiger snake and brown snake, the mulga is one of the most venomous creatures in Australia.

The eastern brown snake (*Pseudonaja textilis*) is extremely dangerous because of its frequent encounters with humans and its aggressive nature. Probably responsible for more cases of snakebite than all the other snakes put together, the eastern brown snake is encountered right down through the farm belt of Queensland, New South Wales, Victoria and South Australia and in central Australia. It averages 1.4 metres and has a cream, yellow or light brown body sometimes with darker blotches, and a uniform coloured back of grey, brown, orange, russet or almost black. Easily aroused, the brown snake becomes extremely aggressive when provoked. It flattens its neck and poises its head in a tensed S-shape, preparing to strike (*opposite and top left*). It attacks high and fast, with an unusual open-mouthed strike, biting several times (*left*). Venom output is low and the fangs are short but the venom is extremely toxic. Diurnal hunters, they feed mainly on mice and lizards and often constrict their prey (as well as using venom).

(*Below*)
Juvenile eastern brown snakes are often banded. Eastern browns, like the taipan, are egg layers, the average clutch numbering twenty to thirty. The young snakes hatch fully formed and receive no attention from the mother snake. Bush fires and predators—eagles, kookaburras, crows, magpies, cats, foxes and other snakes—take their toll and only a small percentage of the thousands of brown snakes hatched every year reach adulthood.

Less aggressive and not as easily aroused as the eastern brown snake, the western brown snake or gwardar (*Pseudonaja nuchalis*) is nonetheless extremely dangerous. It has an extensive range across the drier parts of Australia from Western Australia into the Northern Territory, western Queensland, New South Wales and South Australia. Due to this wide distribution it is one of the country's most abundant snakes, but fortunately it has fairly limited contact with people.

Colour is extremely variable, ranging from yellow to pitch black, with many specimens having black markings around the head and neck. Most have random spots over the whole body and occasionally dark bands. The belly is a yellowish cream with grey or orange or darker blotches. Gwardars average 1.5 metres in length and are oviparous, laying about twenty eggs. Swift-moving and alert, they are mostly diurnal, preying on mice and lizards. One feature of this snake is the surprisingly venomous nature of the juveniles—young gwardars, less than 30 cm long, can cause severe envenomation.

Black snakes are distributed throughout the fertile wetland areas of Queensland, New South Wales, Victoria and south to south-eastern South Australia. These graceful iridescent black snakes are mistakenly believed to be extremely deadly, but are in fact reluctant attackers—they invariably disappear into the undergrowth when approached. If cornered, they flatten out their necks and hiss, but this is usually a bluff. If a bite does occur it is unlikely to prove fatal to an adult, but could be fatal in the case of a child.

The red-bellied black snake (*Pseudechis porphyriacus, top*) prefers wet areas: swamps, lagoons, streams and wet forests. It has a shiny blue-black back with bright red, pink or orange lower sides and a dull red belly. The blue-bellied black snake (*Pseudechis guttatus, above*) is found in the same habitat as the red-bellied, but is also at home in natural grasslands. It usually has a glossy black back with cream spots on some scales and a blue-grey belly, sometimes with yellow spots. Both these snakes grow to an average length of 1.25 metres and feed on small mammals, frogs and lizards. The red-bellied also eats eels and crustaceans and is cannibalistic. The red-bellied produces live young with an average litter of twelve to twenty; the blue-bellied produces eggs, average clutch of about fifteen.

One of the best distinguishing features of the copperhead (*Austrelaps superbus*) is its tolerance of cold climates. They are active earlier in spring and later in autumn than any other species of snake and have even been sighted above the snowline. Only in freezing conditions do they hibernate, making use of animal burrows.

They are found in an arc along the highlands and adjacent slopes of south-eastern Australia, from the South Australia–Victoria border almost as far as the Queensland border, in Tasmania and the Bass Strait Islands. The largest copperheads are found in Victoria, south-eastern South Australia and Tasmania. Copperheads that inhabit the highlands in New South Wales are noticeably shorter and darker in colour, while the specimens found in communities closer to Adelaide are even smaller.

Copperhead venom is highly potent, but these snakes are not easily roused and their strikes often miss. They are heavily built snakes with variable back colour—grey, copper, brown, tan or black. The belly is cream to grey and they are cream, yellow or red along the sides. Young are born live, up to twenty being recorded in one litter. They are usually lighter in colour than the adults. The larger species of copperhead average 1.3 metres. Frogs, tadpoles, other snakes, mice and lizards make up this snake's diet, and where food is plentiful they often congregate in huge colonies.

(*Below*)
One of the most pugnacious and excitable of all Australian snakes—the rough-scaled snake (*Tropidechis carinatus*). This snake is olive green to dark olive in colour with darker bands or splotches. It is sometimes known as the Clarence River snake and is found in cool, moist wetland habitats from northern New South Wales to north Queensland. Their toxic venom has a rapid effect and their fangs are relatively long—5 mm in an average 75 cm specimen. Although not many deaths have been attributed to this snake it is a leading suspect in killings by unidentified snakes in its habitat. As it is mainly nocturnal, hunting for frogs and mice, it is not likely to be seen and identified by anyone bitten by it.

Almost always found in trees, sheltering in tree hollows or behind loose bark, Stephen's banded snake (*Hoplocephalus stephensii*) is ill-tempered when disturbed and attacks eagerly. It is distributed throughout the wet sclerophyll or rainforests of the east coast of northern New South Wales and southern Queensland and is a nocturnal hunter, preying on geckoes, skinks, small birds and mice. These snakes have wide heads and stocky bodies, black backs with narrow cream cross-bands and cream bellies with black blotches towards the tail. Their lips are barred with black and cream. A bite from a Stephen's banded snake can cause acute illness and collapse.

Because of its restricted range, much of which is situated in highly populated areas, the Sydney broad-headed snake (*Hoplocephalus bungaroides*) is becoming increasingly rare. It inhabits a small area within about 250 km radius of Sydney, mainly living under sandstone rocks and sometimes found in hollow trees during summer. It is a stocky snake with a wide head and a black back with narrow yellow cross-bands. Its face is spotted yellow and its belly is grey. Nocturnal, it feeds on lizards and frogs and bears live young in a litter of about twelve. Its venom, which has been likened to that of a southern copperhead, can produce acute symptoms but the snake is not generally regarded as potentially lethal.

De Vis' banded snake (*Denisonia devisii*) occurs in inland Australia from central Queensland to central New South Wales. This heavily built snake grows to half a metre in length and has a light brown back with a series of dark brown bands. The head is dark brown with distinctly barred lips. Nocturnal, it shelters under litter and fallen timber during the day. It has a distinctive defensive stance, holding its body in a series of stiff curves, thrashing about and biting savagely if approached. Although no deaths have been recorded, a bite can produce severe effects and the snake should be regarded as being potentially dangerous.

(*Below*)
A yellowish stripe around the side of the head identifies the golden-crowned snake (*Cacophis squamulosus*). It is dark brown or steely-grey above, with a bright pink belly and black patches and spots on the midline. Nocturnal, feeding mainly on small skinks, this secretive creature shelters under stones, logs and leaf litter during the day. It is distributed throughout the coast and adjacent ranges from central New South Wales to south-eastern Queensland, inhabiting deep forest areas in the north and sandstone areas in the southern part of its range.

Although unlikely to threaten human life, the bite of an eastern small-eyed snake (*Cryptophis nigrescens*) can still be severe. Another secretive, nocturnal snake, the eastern small-eyed is distributed throughout the coast and ranges of eastern Australia from Cape York to southern Victoria. It is usually found under rocks, logs or ground litter or in rocky crevices, and feeds mainly on lizards and frogs. Its back is uniformly black, its belly is cream to bright pink below, often with dark flecks and blotches. Young are born live in litters of from two to five.

(Overleaf)
One of Australia's most attractively coloured venomous snakes, Collett's snake (*Pseudechis collettii*) has striking orange and black markings. It is an uncommon species, restricted to central Queensland. Its venom has been compared to that of the mulga snake and although there is no documentation of lethal attacks, it is a likely killer. Collett's snake averages 1.5 metres in length and preys on rats and mice.

Australia's fastest moving snake, the black whip snake (*Demansia atra, top*) is found on the coastal and adjacent areas of north-eastern and northern Australia from central eastern Queensland to the Kimberleys. These agile snakes have deep narrow heads quite distinct from their thin whip-like bodies. They are usually a rich brown colour, becoming black towards the tail. When the body is dilated, white nicks between some of the scales are visible. Moving only by day, they feed on small lizards and mammals and are normally inoffensive, biting only when handled.

The yellow-faced whip snake (*Demansia psammophis, above*) is found throughout most of continental Australia, apart from the tropical north, from the Kimberleys to Western Queensland. Its colour is variable, ranging from olive green to grey-brown. Whip snakes are so-called because of their rapid lashing motions when disturbed. Sizes vary from 25 cm to 1.5 metres in length—the black whip snake is the largest species. Although not life-threatening, a whip snake bite causes extremely unpleasant symptoms.

Although overseas reports claim human python victims, this is not the case in Australia where the chances of encountering a life-threatening python are minuscule. More realistically, the danger from pythons in this country lies in their bite which is not venomous but can be deep and prone to infection. Pythons' teeth are numerous and back-curved. Arranged in double rows, they are needle-sharp and cause severe lacerations to humans. These reptiles have no poison fangs and kill their victims by first gripping them with their teeth and then coiling their body around them, slowly tightening until the breath is squeezed out of them and they are subdued enough to eat. A python's jaw and head bones are so constructed that the snake can swallow an animal of improbable size. If it catches a particularly large prey it may not need to feed again for a year. They usually prey upon small- to medium-sized mammals and birds.

The largest of Australia's pythons are the amethystine (*Liasis amethystinus*) which averages 3.5 metres and is found in north-eastern Queensland, and the olive python (*Liasis olivaceus*), which averages 2.5 metres in length and is found across northern Australia from Carnarvon, Western Australia, to Cape York.

Diamond pythons (*Morelia spilotes spilotes, above*) and carpet pythons (*Morelia spilotes variegata*) are the most common of all Australia's pythons, being found over most of the mainland except southern Victoria, and most of Western Australia. They are generally nocturnal, feeding on various mammals and birds and often entering barns to forage for rats and mice, providing effective rodent control for farmers. These pythons grow to about 2 metres and are often arboreal, although in many areas they live in burrows made by other animals.

(*Above*)
Australia's largest reptile, the huge saltwater or estuarine crocodile (*Crocodylus porosus*), is a powerful predator and potential man-eater. It is found from the Kimberleys in Western Australia, across the 'Top End' and Gulf of Carpentaria and down nearly the whole of the Queensland coast. The label 'saltwater' is misleading as this crocodile can also be found hundreds of kilometres inland, occupying freshwater rivers and pools. The largest of this lethal species grows up to 6 metres, but the more usual length is from 3–5 metres. Mainly active at night, these crocodiles can be seen basking in the sun on riverbanks during the daytime. Aggressive hunters, they go out at night to feed on fish, waterfowl and animals such as kangaroos and wallabies which they snatch at waterholes. Females lay batches of up to sixty eggs in a carefully constructed nest which they guard until hatching has occurred. This nest is made up of a rotting pile of vegetation, mud and sand—materials which generate a steady heat to incubate the eggs. Relatively few of the fifty-odd eggs that were initially laid hatch after about three months, the baby crocodiles being about 20 cm long.

Drastically depleted by shooting and trapping before protection was granted in the 1970s, estuarine crocodiles are now back with a vengeance and have gained an evil reputation as man-eaters. Since 1975 there have been nine fatal attacks.

The repulsive-looking cane toad has potent poison glands under the warty lumps on his back and can squirt poison from these glands to a distance of 1 metre. This toxin penetrates unbroken skin and causes acceleration of the heartbeat and breathlessness. The frog uses this venom as a defence against any predators. Eating a cane toad—certainly a last resort for humans!—would also cause severe illness as this amphibian's flesh is permeated with poison, particularly in the bulging glands travelling across the shoulders from behind each eye.

This giant toad (*Bufo marinus*), which can reach a length of 25 cm and a weight of almost 1 kg, was introduced to the Queensland canefields in 1935 in an ineffective attempt to control cane beetles. It quickly became a pest itself, invading farms, destroying honey bees and poultry and upsetting the natural ecology by preying on other frogs, insects and spiders. Snakes, cats and dogs have suffered too, dying within an hour of eating the toxin-riddled cane toad.

INDEX

Numerals in *italics* denote photographs